The Giggle Factor (GF). Mention "Planetary Defense" and you'll soon understand. Even without invoking the sinister vision of alien beings arriving to enslave or destroy humanity, the eyebrows of serious and senior members of the national defense and scientific communities go askew when the subject is broached, whether at cocktail parties or congressional budget hearings. Even the most ardent supporters of defending the Earth from cataclysmic cometary or asteroidal impacts share occasional public or private chuckles with colleagues and skeptics—behavior considered unthinkable when discussing means to avert or mitigate the catastrophic epidemics, wars of aggression and genocide, and terrestrial natural disasters that have peppered man's history on Earth.

But the GF doesn't diminish the evidence that the threat exists; measurable and historic, but largely unpredictable. It does, however, continue to cloud a serious issue by generating a unique mosh pit of government branches and agencies, international groups, and private and public research organizations and corporations that would love credit for the resulting system (if it were ever successfully employed), gleefully accept a slice of the multi-billion dollar pie that would accompany a full-fledged planetary defense system, or justify continued funding and development of systems that could be tied to a planetary defense mission. Saddling each of these players, though, are the negative connotations of being involved in a publicly perceived high GF program that may never be used during a human lifetime, raises the suspicions of potential military adversaries, and may not work anyway.

Who should lead? Who should play? Who should pay? The answers are still not clear but the time has come to stop the giggling, state the problem, sort out the mosh, and take some serious steps toward providing for the common defense—of the planet.

The threat is real. Strikes from space presented in broadcast and print news over the last decade, in scholarly journals and books from the last century, as well as in the geological record document asteroid and comet strikes. Several recently published books and papers present overwhelming evidence of large asteroid strikes dating back millions of years that have had a profound impact on the evolution of our planet. An object impacting the Earth 214 million years ago caused the 100 kilometer-wide Manicouagan Crater in Quebec, while 65 million years ago an asteroid estimated to have a diameter of 10 kilometers caused a 180 km crater off the coast of the Yucatan Peninsula, and is today credited with ridding the world of the dinosaur menace.[1] Since the discovery of these and other craters around the world, several researchers have determined a periodicity for 10 km-class asteroid impacts of 26-30 million years dating back over 600 million years, with a less distinct recurrence of larger objects.[2] The periodicity of 10 km-size asteroid strikes correlates reasonably well (within the error range of geologic dating) with major global extinction events (as many as ten) recurring on a 26 million year cycle dating back 260 million years. Fortunately, the last of these events (Middle Miocene) occurred about 16 million years ago![3]

Safely distant in time from the next major extinction cycle—theories for the cause of which are just slightly beyond the scope of this paper—an examination of more recent impacts and near misses yields reason for concern. The poster child for planetary defense is undoubtedly the Tunguska Event, named for the very sparsely populated Siberian region over which a 100,000 metric ton (estimated) asteroid exploded at an altitude of six kilometers on 30 June 1908 with the energy of 2,000 Hiroshima-sized nuclear weapons. Trees were felled over 2,150 square kilometers, with resulting fires wiping out twice that area.[4] The shock wave caused by the explosion circumnavigated the globe, and was recorded in Potsdam, Germany on successive days.[5]

While the last confirmed, widespread loss of life caused by a cosmic impact occurred in China in 1490 (an estimated 10,000 died), a "mini-Tunguska" occurred over New Guinea in 1994 with resulting blast energy of 11 kilotons.[6] Any of the nine large cometary fragments (designated Shoemaker-Levy 9) whose spectacular Jovian impacts we witnessed in 1994, would have caused tremendous damage on Earth, far in excess of that seen in Tunguska. Two near misses (asteroids 1989FC and Toutatis, in 1989 and 1992, respectively) also highlighted our vulnerability. Toutatis, with its 4 km diameter, possessed the energy of 9 million megatons of TNT while passing less than two lunar distances of Earth,[7] and 1989FC passed at a similar distance with no warning—having never before been catalogued!

As a final note in an effort to highlight the threat posed by asteroids of all sizes, one need only look back a few months and a bit north to the Yukon Territory of Canada. On 18 January 2000, a small meteor (estimated at several kg) entered the Earth's atmosphere and exploded at an altitude of about 25 km. While the explosion (equivalent to between two and three kilotons of TNT) shook houses and was witnessed over an area of thousands of square miles in this sparsely populated region, the most interesting effect surprised many observers.[8] It seems that the meteor's explosion produced an electromagnetic pulse similar to that of a low-yield nuclear device—an effect of nuclear weapons known to have dire consequences for electronic equipment and often predicted as a precursor to a nuclear strike curing the Cold War.

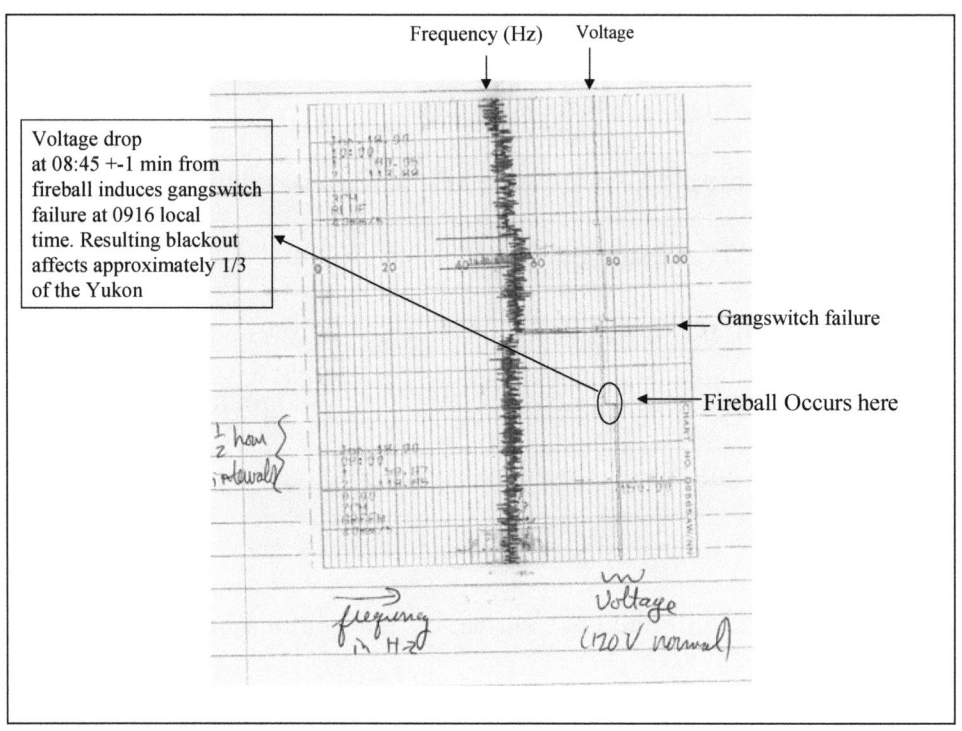

Figure 1 Yukon Power Grid Voltage, 18 Jan 00

Figure 1 above shows the voltage spike measured in the (admittedly small) Yukon power grid.[9] This spike, in turn, caused a power outage over one-third of the province with power restored some hours later. In imagining a similar incident occurring over a major metropolitan area, the possibilities for damage, panic and misinterpretation seem significant. A meteor of this size may not be large enough to identify far enough in advance to divert it (and the cost to destroy or divert it may not justify such an operation), but its timely detection and the subsequent warning of its expected strike could save many lives and reduce property damage greatly. The event also serves as another vivid reminder of the frequency with which meteor and asteroid reentries with measurable effects occur.

The current estimate of Near-Earth Objects (NEOs) larger than one kilometer in diameter is about 2,000, based on confirmed sightings of just over 200 and the amount of space that remains to be fully searched. Based on this information and other data

4

concerning the frequency of large-scale impacts on the Moon and from regions of the globe with more complete historical record of impacts (e.g., Australia, where the sparse vegetation and arid climate combine to limit weathering of craters), scientists estimate the chance of dying from an asteroid or comet impact in the United States is 1 in 20,000.[10] By comparison, the chance of dying in an aircraft accident is also about 1 in 20,000, dying in a tornado is a 1 in 60,000 shot and death by venomous bites or stings is 1 in 100,000. You are, however, more likely to die in a car accident (1 in 100), be murdered (1 in 300) or electrocuted (1 in 5,000) than die as a result of a cosmic impact.[11]

Perhaps the greatest intellectual challenge in dealing with this threat is the extraordinarily low annual likelihood of occurrence coupled with the incomparably dire consequences.[12] The higher than expected likelihood of "death by asteroid" is attributed to the supposition that no event short of global nuclear war has the potential to kill tens or hundreds of millions of people—more than accounting for a large-scale impact's low chance of occurrence. There is, however, no "relevant history" for an asteroid strike causing a global catastrophic event, so even if it is inevitable that such a strike will someday occur, few are able to internalize the risk or view a need for action. Political changes need constituencies and 'people who will be harmed by an impact' simply do no make up an identifiable constituency today[13]--unlike the millions who fight for funding to further diminish those threats which are, statistically, far less likely to kill them (e.g., nuclear power accidents).

Before any discussion of mitigating the threat posed by asteroids and comets is undertaken, these objects must be found. The first active step, then, in implementing any planetary defense system must be the effective surveillance of space—not only to track known objects whose trajectory may change with time, but also to identify and catalogue threatening NEOs in Earth-crossing trajectories, and "one time" asteroids and comets coming from much further out in the Solar System.

Prior to 1990, astronomers identified asteroids haphazardly by searching the Asteroid Belt between Mars and Jupiter, roughly 200,000,000 km distant. Relatively small and faint, the light reflected by a one-kilometer object at that distance equates to a stellar magnitude of about 22,[i] requiring a sophisticated, ground-based two meter aperture telescope to be identified and tracked. Without any organized plan to catalogue the skies, experts estimated that several centuries would pass before all potential Earth-crossing asteroids (ECAs) would be identified—based on the amount of sky already surveyed, the search pace, and the number of objects detected and predicted.[14]

In 1990, though, amid concern over the cosmic impact hazard caused by 1989FC and the inevitable media attention it generated, Congress directed NASA to conduct two workshops to discuss means of dealing with the NEO threat—one on detection, the other on interception—as part of the NASA multiyear authorization act. The goal of the detection workshop, held in January 1992, was to accelerate ECA identification, completing a census in less than 25 years. The detection workshop's product, a plan named the Spaceguard Survey, borrowed the name of a similar project described by science-fiction author Sir Arthur C. Clarke 20 years earlier in his novel, *Rendezvous with Rama*.[15]

NASA officials briefed Spaceguard to Congress on 24 March 1993. The chairs of the two workshops mentioned earlier, David Morrison (detection) and John Rather

---

[i] The limit of unaided human vision is magnitude 6, while magnitude 22 is more than one million times fainter

(interception), accompanied Wesley Huntress, NASA's Associate Administrator for Space Sciences, testified before the Subcommittee on Space of the House Committee on Science, Space, and Technology. The detection workshop's recommendations for Spaceguard centered on the identification of NEOs one kilometer in diameter or greater, based on one to two kilometer asteroids representing the threshold at which a global catastrophe might be induced. The relatively large size of these objects also means that they can be located using ground-based telescopes, as opposed to more expensive space-based systems that would take many more years to develop and deploy.[16] To detect these objects, Morrison and the members of the Detection Workshop, proposed a network of six specially constructed telescopes, each with an aperture of two to three meters—intermediate size by current astronomical standards—equipped with modern, large-format electronic detectors. The use of new electronic techniques is especially critical to decrease the time needed to complete the catalogue, since prior efforts were based on labor-intensive analysis of photographic film.[17] Such a system would also be able to identify previously unknown comets by the time they were near Jupiter's orbit, providing at least two years warning before Earth passage or strike—hopefully enough time to construct and launch a mission to divert or destroy the threat.

To accomplish the identification task within the projected time period, Morrison's group estimated a total cost of about $300M over 25 years (1993 constant dollars), with $50M allotted to capital investment (six telescopes at $6M-$8M apiece plus other infrastructure) and $10M-$15M per year in operations costs. They recommended that this funding be shared among many nations—especially those such as Australia who would play host to a surveying telescope—and that additional NASA funding to cover the program begin in fiscal year 1993 with $3M, and totaling $49M between FY93 and FY00.[18] Unfortunately, little was done in response to the Spaceguard report. No new telescopes were purchased and NASA only made a small amount of money available—much less than a $1M, which was barely enough to keep the ongoing, baseline effort in the U.S. alive.[19]

The press, and as a result Congress, were again energized to the NEO threat during 1994 when the spectacular Shoemaker-Levy 9 comet fragments hit Jupiter and were displayed in real-time on the internet and television. Hearings were held on Capitol Hill and determined congressmen again took bold action by commissioning yet another committee to recommend courses of action. Dr. Gene Shoemaker led the Near-Earth Object Study Working Group, whose product is today rightfully called the "Shoemaker Report."[20]

Figure 2 Shoemaker-Levy 9 "Fragment G" Strikes Jupiter, 18 July 1994[21]

The Shoemaker Report recommended a more ambitious goal of finding 90% of the estimated number near-Earth asteroids and short period comets larger than 1 kilometer within 15 years, or within 10 years if the U.S. Air Force and international agencies employed their assets to significantly augment NASA's efforts. The working group estimated the cost of such a program to be about $60M over 15 years (1995 constant dollars), including $24M in start-up costs over the first five years to purchase two 2-meter aperture telescopes and equip several new and existing 1-meter telescopes with the state-of-the-art electronic detectors and software described earlier.[22]

A key component of the Shoemaker Report, as in the earlier Spaceguard Survey, was its international character. However, it seems that most nations interested in the NEO threat are still awaiting America's lead. Russia, for example, has the technology and interest (Tunguska) among its astronomy and military communities to play a significant role in the Spaceguard Survey, but economic circumstances have precluded them from taking the initiative. Australia has recently backed away from its fledgling telescope program, which played a critical role in confirming NEOs first seen by other telescopes from its unique location in the southern hemisphere, and international attempts to encourage the Australian government to bring its program back into operation have failed.[23] The United Kingdom, home of some of the most enthusiastic NEO watchers, formed a "Task Force on NEOs" led by Dr. Harry Atkinson. This group of four scientists has limited funding and is only tasked with making recommendation to Her Majesty's Government by mid-2000 on how the UK should best contribute to the international effort on NEOs.[24] Additionally, Spaceguard is a loose, voluntary consortium of international observatories and interested parties that serves to relay NEO identification to concerned groups and fellow participants.

The latest round of testimony before Congress on the subject of NEOs was held on May 21, 1998. The opening statement by the Honorable Dana Rohrbacher, Chairman of the Subcommittee on Space and Astronautics, makes it clear that his reason for calling this hearing on Space Day was not purely bipartisan. Long a supporter of NEO detection and mitigation, as well as other civil and military space programs, Rep Rohrbacher was disturbed that NASA had not committed funds for NEO detection in line with the Shoemaker Report's recommendations. NASA apparently exacerbated his displeasure by reprogramming $50M to fund Vice President Gore's idea of building and launching a satellite that would "sit" at an Earth-Sun lagrangian point[ii] and simply beam a constant view of the Earth down to internet users everywhere. Rep Rohrbacher suggested that the Vice President's "harebrained idea" might prove useful "if we'd like to watch from a distance as the Earth is being pulverized by asteroids."[25] The hearing was also motivated by the President's Fall 1997 line-item veto of funding for the Air Force's Clementine II asteroid interceptor demonstrator,[iii] as well as the impending release of the movies *Deep Impact* and *Armageddon*.

The hearings of 21 May 1998 constituted a "status check" on progress in identifying NEOs and methods to mitigate their threat following the aforementioned Spaceguard Survey and Shoemaker Report. Testifying before the committee were

---

[ii] A lagrangian point in space is where the gravitational pull of two large bodies (the Earth and Sun in this case) on an object "balance" resulting a constant relative position of the subject spacecraft relative to the two larger bodies.

[iii] Many in the administration viewed Clementine II as a thinly veiled prototype for a space-based missile interceptor—a not entirely inaccurate view since its mission was to demonstrate microsatellite technology, "cold body" sensors, low-cost propulsion and guidance systems common to potential anti-ballistic missile systems like "Brilliant Pebbles."

representatives from NASA, academia, national research labs and federally funded research and development corporations. Despite the fact that much of the progress in NEO detection is directly attributable to Department of Defense (DoD) cooperation, no military panelists were invited to the hearings. Members of the panel widely acknowledged military involvement in NEO detection, though, and suggestions for further leveraging military space surveillance capabilities played a large role in the hearings.

In the years since the Shoemaker Report was published, several sources suggested that the DoD, specifically the U.S. Air Force, play a much larger role in the search for NEOs. Bell *et al* and Chapman, among others, recommended using existing Air Force 1m aperture telescopes from the Ground-based Electro-Optical Deep Space Surveillance System (GEODSS).[26,27] The first use of a GEODSS telescope under the joint Air Force-Jet Propulsion Laboratory (JPL) Near Earth Asteroid Tracking (NEAT) program began on the rim of the Haleakala Crater in Maui in 1996. However, the Air Force cut time allotted to the NEAT mission to six nights per month in early 1998, drastically reducing its possible contribution to NEO identification.[28] Since that time, the Air Force removed the NEAT camera in September 1998 to complete the Air Force's GEODSS Modernization Program and has since reinstalled it on the Air Force Research Laboratory's 1.2-meter aperture telescope on Maui. Routine NEO detection operations have since resumed.[29]

The Air Force also funded development of the Lincoln Near Earth Asteroid Research (LINEAR) program as a prototype next-generation GEODSS detector. A single LINEAR system began operations in early 1998 in Socorro, New Mexico[30] and according

11

to the testimony of Dr. Carl Pilcher of NASA, his organization is now prepared to take over LINEAR program funding.[31] Currently, LINEAR is responsible for about half of the NEO scanning task worldwide—about 8,000 square degrees of sky per month.[32] This corresponds to 40% of the scanning goal set by the Spaceguard Survey and Shoemaker Report and is the primary reason the detection rate of 1 km and larger NEOs soared recently. Prior to publication of the Shoemaker Report, 188 NEOs of this class had been located. Between July 1995 and April 1998, NEO searching telescopes identified only 40 more. Since then, however, over 110 new NEOs were identified.[33] In fact, Dr. Pilcher now believes that NASA and its domestic and international partners are well on the road to achieving the goals of the Shoemaker report—identifying 90% of NEOs larger than 1 km within the next 10 years. NASA now seems ready to fund the detection programs at the level directed by Congress, though there remains some uncertainty as to what organization should ultimately coordinate the effort.[34]

Candidates to act as coordinators of NEO search and cataloguing efforts range from U.S. governmental organizations such as NASA and United States Space Command to international bodies like the International Geophysical Union (IGU), the International Astronomical Union and the United Nations Space Committee. The Federal Emergency Management Agency (FEMA) has, to date, played no role whatsoever on the NEO issue.[iv] The IGU currently gathers inputs on NEOs from about 30 different countries, but has no real funding of it own. Dr. Pilcher, during his congressional testimony, recommended bringing all of the above-named bodies together, along with other interested parties to submit a plan to the U.N. for endorsement and participation.[35]

[iv] FEMA makes no mention of the threat from space in its 10-year strategic plan, entitled "Partnership for a Safer Future," available on-line at http://166.112.200.140/library/spln_1_htm

Others doubt such a loose confederation could effectively lead such an ambitious program, particularly if NEO hunters identified an actual threat to the Earth and action was needed to divert or destroy the asteroid.

The leading advocate NEO detection and mitigation within the DoD is Brigadier General (select) Simon P. "Pete" Worden, currently serving as the Deputy Director for Command and Control (XOC), reporting to the Deputy Chief of Staff of the Air Force for Air and Space Operations, at the Pentagon. Worden, trained as a research astronomer with a PhD from the University of Arizona, is probably best known in the space and astronomy community as director of the 1994 lunar probe named Clementine that orbited the Moon and found some of the first evidence of water possibly hidden in the soil of polar craters there. Clementine II, mentioned earlier, planned to rendezvous with an asteroid until President Clinton's line-item veto killed the program. During a series of assignments at the Ballistic Missile Defense Office (BMDO), the White House Staff, Air Force Space Command and the Air Staff, Worden championed the use of GEODSS and development of LINEAR to catalogue 1 km-class asteroids, as well as the exploitation of microsatellite technologies to identify smaller NEOs and as a low-cost method to divert Earth-threatening asteroids.

On 7 February 2000, Worden ignited a minor firestorm within the NEO community when he submitted an essay proposing DoD leadership of international efforts to detect, study and (if necessary) defend against NEOs to the Cambridge-Conference Network electronic newsgroup.[v] Not meant as an official policy statement of the U.S government or the Department of the Air Force, Worden's personal view was that while identification of 1 km asteroids seemed to be progressing at a sufficiently rapid (indeed, accelerating) pace, governments and astronomers were neglecting the 100 meter or "Tunguska Class" objects that strike up to several times per century. Worden couched the challenge in terms of space situational awareness, noting that the U.S. space community, particularly the DoD and NASA, are beginning to understand the importance of identifying and tracking virtually everything in Earth orbit in order to protect peaceful operations there, now and in the future. The ground and space-based tools used to achieve this close-range awareness would also serve to greatly improve detection of NEOs much smaller than mentioned by either Spaceguard or Shoemaker, but still capable of causing considerable death and destruction on Earth.[36]

Worden recognizes that the preponderance of NEO detection capability is American and recommends that other nations not spend their money duplicating the U.S.'s network of sensors. Rather than spend the hundreds of millions of dollars on telescopes and sensors, he claims that simple economics argue that this portion of the NEO problem be ceded to the U.S. military, though he acknowledges not everyone in the DoD is as eager to take on the NEO task!

Instead, European and other countries could play to their strength by developing 100 kg-class microsatellites costing about $5M-10M to build and an equivalent amount to launch. Several European groups lead the world in the development of these microsatellites (e.g. the University of Surrey in the UK), thanks in no small measure to the ability of the European Space Agency's Ariane IV and V launchers to carry up to

---

[v] The Cambridge Conference Network (CCNet) is a scholarly electronic network moderated and edited by Professor Benny J. Peiser of Cambridge University, UK, that focuses on the link between asteroid and cometary impacts and geological evolution.

eight microsatellites as auxiliary payloads at a cost of about $1M per satellite.[vi] NASA, U.S. academic institutions, and U.S. aerospace companies have begun developing microsatellites for space science and planetary exploration missions as well, while the DoD is beginning to develop microsatellites for servicing and refueling larger mission satellites. The combination of these domestic and international systems should allow for a wide range of low-cost missions to NEOs—including missions to fully characterize their structure and possibly bring back samples to Earth. More advanced microsatellites could later assist in the surveillance and cataloguing of NEOs as some doubt remains whether existing ground-based sensors based on the Earth can detect all of the potentially threatening objects.[37]

Worden recommends these *in situ* studies of NEOs using low-cost microsatellites begin immediately and should involve NASA, ESA, other nation's space agencies as well as the DoD, and use the latest technology to rendezvous, inspect, sample, and even impact NEOs to study their composition and structure. An estimated cost of about $10M-$20M per mission, including data reduction and launch, makes this a far more affordable prospect than the $60M-$200M cost of dedicated, full-sized satellite missions. He goes on to cite the UK's recent decision to stand up a formal NEO task force, mentioned earlier, and significant contributions by Canadian academics to support study of the Leonid meteor storms of 1998 and 1999 as evidence of official interest in NEOs and willingness to participate in more active programs.[38]

Worden continues by suggesting that the time and money needed to develop mitigation systems against NEOs today argue against such a move until a bona fide threat

---

[vi] United States Air Force is putting a similar auxiliary microsatellite adaptor on our new EELV (Evolved Expendable Launch Vehicle) launch systems.

emerges. This would also avoid much of the political consternation arising from some nuclear weapons experts who advocate weapons retention and even testing in space. Such testing would be of questionable value since we can't reliably divert an NEO until we know much more about its structure. A decade of dedicated microsatellite missions to characterize the threat and best devise means of mitigation (including impacting some small NEOs as a side experiment) would yield the necessary experience should a true threat emerge.[39]

In fact, NASA and other organizations have already sent spacecraft to investigate comets and asteroids, though at considerably more cost and less frequency than Worden proposes with his microsatellites.[vii] The NEAR (Near-Earth Asteroid Rendezvous) satellite passed by the asteroid Mathilde in 1998 and rendezvoused with Eros earlier this year. NEAR is currently in orbit around Eros and though it will not return to Earth with samples from the asteroid, it is conducting observations and studies from a distance of the object that is about twice the size of Manhattan Island. In fact, characteristics of Mathilde and Eros already discovered by NEAR highlight potential problems we might face should we need to divert an asteroid in the future. Mathilde's density is only 30% greater than that of water and it's marked with some unexpectedly large craters suggesting the asteroid contains internal voids and is capable of absorbing tremendous amounts of energy without fragmenting. Eros is somewhat smaller, but denser than Mathilde and also has sizeable craters (see Figure 3).

[vii] The appeal of microsatellites lies partially in the fact that every time a communications satellite is launched into a geosynchronous orbit (which occurs over a dozen times per year), up to eight 100kg microsats can "piggyback" on-board and be delivered into a high-energy geosynchronous transfer orbit, from which the escape of Earth's gravity is possible with reasonable amounts of fuel.

Figure 3 Asteroid Eros Taken from NEAR Spacecraft in 200 km Orbit[40]

The idea that the Air Force has a lead role to play in NEO detection and

mitigation is supported half-heartedly, if at all, by the corporate Air Force. Whereas no

mention of planetary defense can be found in USSPACECOM's Long Range Vision,[41]

the ability to "detect Earth-crossing objects" is listed among Air Force Space Commands

deficiencies. It is, however, ranked at the bottom of Space Command's Space Control

needs and 62nd, 54th, and 53rd overall of the command's overall 65 near-term, mid-term

and far-term prioritized deficiencies, respectively.[42] A recent article in the Air Force's

Aerospace Power Journal by Lieutenant Colonel Cynthia McKinley does suggest that the

role for space surveillance and debris and asteroid mitigation be given to a "Space

Guard," as she would name a separate service organized to patrol and protect assets in

space much as the U.S. Coast Guard performs at sea.[43] This view, though, is not a

reflection of official Air Force policy and is unlikely to become so in the foreseeable

future.

The scientific community is not united behind Worden's proposals either. Though many find merit in his thoughts, others (including some CCNet subscribers) question whether defining NEO detection as a military mission might inhibit scientific and international cooperation or might again be a convenient cover for developing technologies critical to a national missile defense system, again raising the specter of weaponizing space.

Where then are we left? BGen (sel) Worden's essay makes an excellent case for the priority of detection and characterization over mitigation efforts. If we cannot organize and execute a coherent plan to accomplish these first two tasks, then it would certainly seem foolish to establish a funded program centered on mitigation. The GF alone argues for concretization of the threat (both physically and in the public eye) prior to manning battle stations. Congress and the American public clearly see NASA as the lead agency for the moment based on congressional testimony, funding, and on-going and future space programs. NASA, based on congressional testimony, wants just as much to maintain funding control and oversight, though Congress doubts NASA's commitment to those roles for both partisan political reasons and historical record of past lukewarm support.

The Air Force, however, commands the lion's share of assets needed to complete the identification task, both on the ground and currently in orbit, and has led funding and research for future systems such as LINEAR. The fact that future operational Air Force systems such as the Space-Based Infrared Radar System (SBIRS) could augment the search for NEOs using on-board sensors and the latest Air Force launch vehicle, the Evolved Expendable Launch Vehicle (EELV), is being adapted to carry secondary microsatellite payloads Ariane-style argues strongly for an increased Air Force role. The corporate Air Force leadership, though, has not publicly stepped up to claim the mission and seems hesitant to do so for fear it may be forced to redirect already scarce resources to a job many see as far-removed from that of traditional warfighter.

The global nature of the threat, meanwhile, and the interest of several other nations, suggest an international organization take the lead. NASA officials suggested a U.N. body endorse any world-wide plan for NEO mitigation and lobby for international participation, but members of Congress and the DoD may be hesitant to include some of the potential interested nations in a program containing ties to high-technology, defense-related systems. Also, few nations have expressed much more than academic interest in planetary defense—eschewing actively funded hardware programs in the near-term for study groups, conferences, and task forces to make recommendations for future action.

Weighing all of these factors leads me to suggest, in the near-term, the following actions should be taken to address the NEO problem adequately:

1. The President should direct, through the Secretary of Defense, that detection and mitigation of natural threats from space is a military mission assigned to U.S. Space Command. Title 10 of the U.S. Code (Armed Forces) should then be amended by Congress to reflect CINCSPACE's new responsibility and the fact that space is now an AOR (Area of Responsibility) in which threats exist and operations might arise whereby USCINCSPACE will be the "supported CINC." Air Force Space Command (specifically, 14th AF) should be designated as the component command responsible for executing the NEO mission.

2. NASA should assign an associate administrator with appropriate staffing to Air Force Space Command as deputy director for accomplishing the NEO mission. In this role, the NASA designee should act as primary liaison

between NASA's space sciences program (and related academic and research interests) and Air Force Space Command to ensure proper planning, funding and technical direction.

3. Congress should consolidate all funding for the NEO mission within the Air Force (transferring current NASA funding to the DoD), with oversight provided by the House Committee on Space and Aeronautics. The NASA Administrator and USCINCSPACE should act as advocates for the program before the Committee.

4. Liaisons from interested foreign nations should be invited to participate in the program, and be located Air Force Space Command Headquarters, to ensure integration of their nation's contributions into the overall effort. Supporting nations should have representatives on the planning and operational staffs much as Canadian officers now serve on the NORAD staff.

5. Detection efforts should continue per the Shoemaker Report with additional funding to support LINEAR and the augmentation of all remaining GEODSS telescopes with NEAT systems. International partners willing to house and operate GEODSS telescopes should be encouraged to do so.

6. Planning for a coalition space mission using one or more microsatellites to rendezvous with and analyze a 1 km-class asteroid should begin immediately. Plans for a series of such missions should be aimed at characterizing an adequate sample size (~10) of asteroids within a decade in order to begin work on mitigation systems.

7. FEMA should be directed to include asteroid and cometary impacts in their planning, as well as nominating a representative to serve as liaison to the Air Force Space Command organization responsible for NEO detection, tracking and mitigation.

Though implementation of these steps may cause a few giggles inside the long hallways of the Pentagon and other locations inside the Beltway, they seem to be most prudent in terms of the threat we face and the players best able to address the problem.

For anyone who has seen Meteor Crater in Arizona or been awakened at night by an exploding meteor,[viii] these steps seem prudent and quite economical. There is no reason they should cost more than $500M over the next 10 years—15 space missions at a total cost of $20M apiece, plus operational and capital costs for telescopes—in line with Spaceguard and the Shoemaker Report. Detection is the key to the problem today, since it affords us the time to act. If our national leadership can act responsibly to address this problem, instead of reacting to close-calls, partisan politics, media hype, and movie releases then we may well lay the groundwork for an effective means of defending our planet from space threats that once caused giggles.

U.S. Representative George E, Brown, Jr. (D-CA) said:

If some day in the future we discover well in advance that an asteroid that is big enough to cause a mass extinction is going to hit the Earth, and then we alter the course of that asteroid so that is does not hit us, it will be one of the most important accomplishments in all of human history.

Taking the steps mentioned above could play an important role if we are ever faced with this challenge, because giggling certainly won't do the trick.

[viii] It happened to me and half the city of Colorado Springs, Colorado in January, 1998.

ENDNOTES

[1] Rosario Nici and Douglas Kaupa, "Planetary Defense: Department of Defense Cost for the Detection, Exploration, and Rendezvous Mission of Near-Earth Objects," *Airpower Journal* XI, no. 2 (Summer 1997), p. 94.

[2] Duncan Steel, *Rogue Asteroids and Doomsday Comets*, John Wiley & Sons, New York, 1995, pp. 95-96.

[3] J.J. Stepkoski, "The Taxonomic Structures of Periodic Extinctions," *Geologic Society of America*, Special Paper 247, 1990, pp. 33-44.

[4] Roy A. Gallant, "Journey to Tunguska," *Sky and Telescope* 87, No. 6 (June 1994), p. 38.

[5] Larry D. Bell, William Bender, and Michael Carey, "Planetary Asteroid Defense Study: Assessing the Risk and Responding to the Natural Space Debris Threat," research paper ACSC/DR/225/95-04 (Maxwell AFB, AL: Air Command and Staff College, 1995), p. 58.

[6] John M. Urias et al., "Planetary Defense: Catastrophic Health Insurance for Planet Earth," research paper submitted to Air Force 2025, October 1996, http://www.au.af mil/au/2025volume3/chap16/.

[7] John C. Kunich, "Planetary Defense: The Legality of Global Survival," *The Air Force Law Review* 41, 1997, p. 119.

[8] Ron Baalke baalke@jpl.nasa.gov, "DoD Fireball Detection Over Yukon Territory, Canada," CCNet 07/2000, 20 January 2000.

[9] Personal e-Mail from Dr. Peter Brown, Meteor Physics Lab, Department of Physics and Astronomy, University of Western Ontario, London, Ontario, N6A 3K7, Canada, 1 March 2000.

[10] Clark R. Chapman and David Morrison, "Impacts on Earth by Asteroids and Comets: Assessing the Hazard," *Nature* 367, No. 6458, 6 Jan 94, pp. 33-40.

[11] Ibid.

[12] George J. Friedman, "Risk Management Applied to Planetary Defense," *IEEE Transactions on Aerospace and Electronic Systems* 33, No. 2 (April 1997), pp. 722-723.

[13] NEO News, NASA Ames Research Center, 10 February 1998, cited by James S. Knox, "Planetary Defense: Legacy for a Certain Future," Air War College Research Paper, April 1998.

[14] "The Spaceguard Survey: Report of the NASA International Near-Earth-Object Detection Workshop," David Morrison, ed., 25 January 1992, Executive Summary, pp. v-vi.

[15] Ibid., Chapter 9, p. 49.

[16] David Morrison, statement before the Subcommittee on Space of the Committee on Science, Space, and Technology, U.S. House of Representatives, March 24, 1993, p. 13.

[17] Ibid.

[18] "The Spaceguard Survey: Report of the NASA International Near-Earth-Object Detection Workshop," David Morrison, ed., 25 January 1992, Chapter 9, p. 52.

[19] Steel, p. 220.

[20] Clark R. Chapman, "Statement on the Threat of Impact by Near-Earth Asteroids," before the Subcommittee on Space and Aeronautics of the Committee on Science of the U.S. House of Representatives, 21 May 1998, p. 5.

[21] http://nssdc.gsfc.nasa.gov/sl9/image/sl9g_hst5.gif

[22] Ibid., pp. 5-6.

[23] Ibid., pp. 6-7.

[24] Personal e-Mail communication with Dr. Harry Atkinson, Chairman, United Kingdom NEO Task Force.

[25] The Honorable Dana Rohrbacher, opening statement, "The Threat and the Opportunity of Asteroids and Other Near-Earth Objects," Hearing before the Subcommittee on Space and Aeronautics of the Committee on Science of the U.S. House of Representatives, 21 May 1998, pp. 1-2.

[26] Bell et al, pp. 196-215.

[27] Chapman, pp. 8-9.

[28] Ibid., p. 7.

[29] Lindley Johnson, "Near Earth Objects Team Report" (draft), component briefing of the next Air Force Space Command/NASA/NRO Partnership Council Meeting, date to be determined.

[30] Carl Pilcher, testimony before the Subcommittee on Space and Aeronautics of the Committee on Science of the U.S. House of Representatives, 21 May 1998, p. 37.

[31] Ibid., p. 42.

[32] Johnson

[33] Ibid.

[34] Pilcher, p. 44.

[35] Gregory Canavan and Carl Pilcher, testimony before the Subcommittee on Space and Aeronautics of the Committee on Science of the U.S. House of Representatives, 21 May 1998, pp. 84-85.

[36] Simon P. Worden, "NEOs, Planetary Defense and Government—A View From the Pentagon," CCNet-Essay, http://abob.libs.uga.edu/bobk/ccc/ce020700.html

[37] Ibid.

[38] Ibid.

[39] Ibid.

[40] www nasa.gov

[41] "Long Range Plan: Implementing USSPACECOM Vision for 2020," U.S. Space Command, March 1998.

[42] "1998 Strategic Master Plan," Air Force Space Command, 20 March 1999, pp. 33-34.

[43] Cynthia J. McKinley, "The Guardians of Space: Organizing America's Space Assets for the Twenty-First Century," *Aerospace Power Journal* 14, No. 1, Spring 2000, pp. 37-45.